S0-ARL-826

SCIENCE IN ACTION

Robots

PETER MARSH

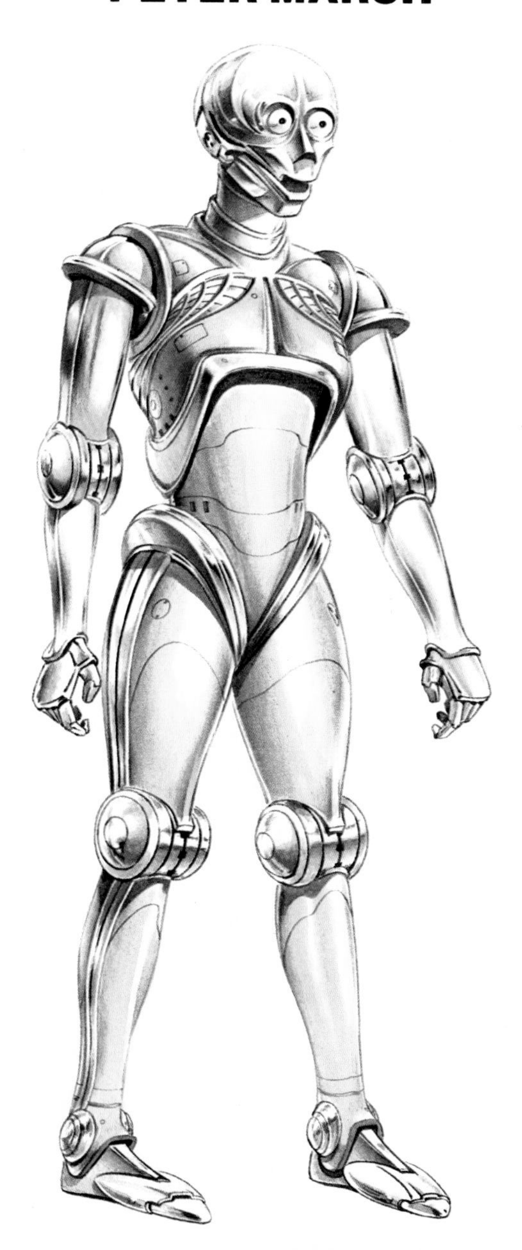

WARWICK PRESS

Published 1983 by Warwick Press,
387 Park Avenue South, New York, New York 10016.

First published in Great Britain by Kingfisher
Books Limited, 1983.

Printed in Italy by Vallardi Industrie Grafiche, Milan.

6 5 4 3 2 1

ISBN 0-531-09223-2
Library of Congress Catalog Card No. 83-50343

Contents

Introducing Robots

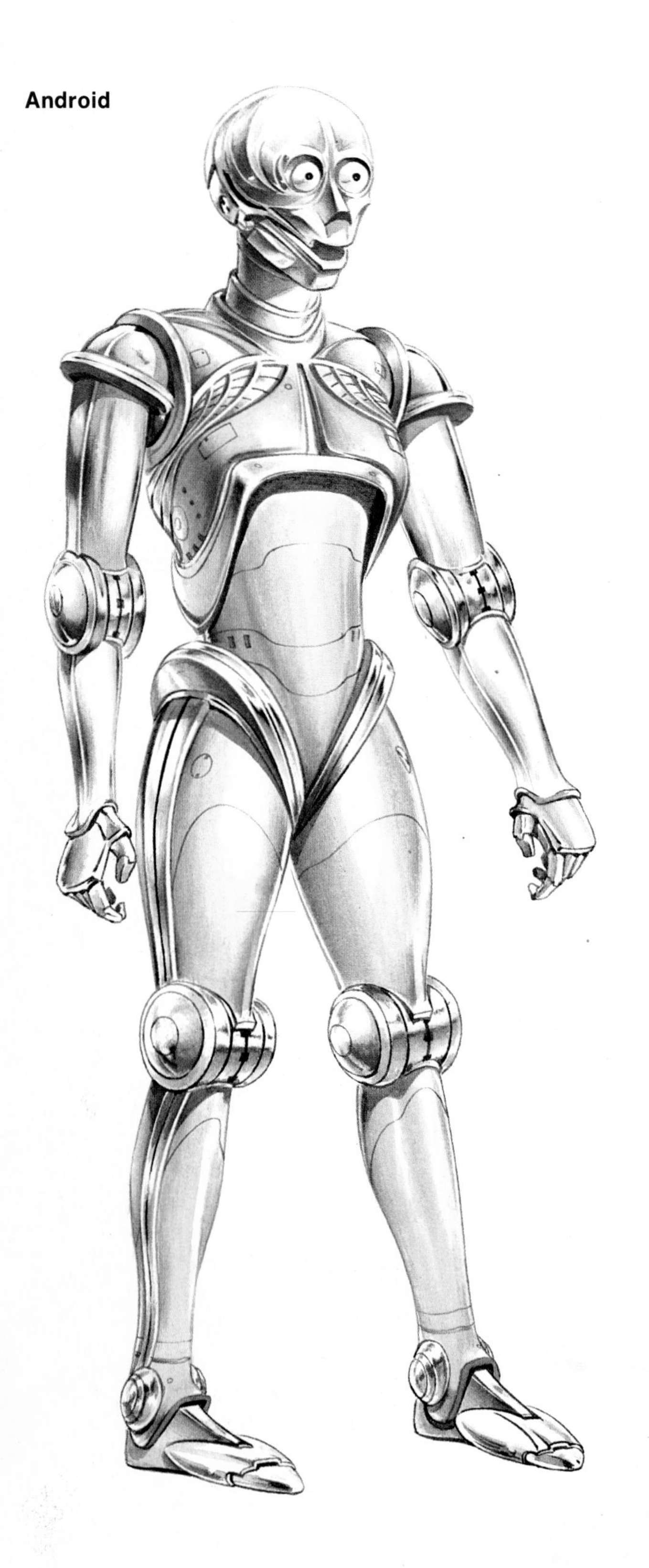

For many years robots have excited people's imaginations – whether in books, in films or inside fairgrounds or amusement arcades. But it is only quite recently that robots have left the realms of fiction. They are becoming useful to men and women in all sorts of ways. How this change has come about, and how it will progress over the next few years, is the subject of this book.

Even now thousands of robots are at work around the world. Their activities range from delicate construction work to heaving huge loads, and they can be found anywhere from factories and nuclear power stations to outer space. In the future robots may become an even greater part of our everyday lives. With the help of "brains" that are in fact powerful computers, robots may become as clever as people. They could take over a wide variety of jobs – in offices, in factories or in the home. Robots could also work in particularly unpleasant places where complex, vital work needs to be done, such as mines or the bottom of the ocean.

What is a Robot?

But first, what exactly do we mean by "robot"? The word can be defined differently depending on who you talk to, or in which part of the world you happen to live. The widest possible definition of a robot is "a machine which imitates a person in appearance or action". Note the emphasis on some kind of physical likeness to people. Over the years, robot-makers have been especially interested in designing robots which, in some way, look like men or women. It is this

◀ For many years people thought of robots as devices that looked and behaved like humans – androids or mechanical men. Nowadays robots have been redefined to mean mechanical arms that are controlled by a computer.

▶ One of the best-selling industrial robots is the Unimate, made by the American firm Unimation. This machine can do a variety of jobs depending on how it is controlled. It can weld metal or move objects from place to place.

hich has given robots their superhuman repu-
ıtion. People regard them with a mixture of
ıscination and fear. The idea that a machine
ın look like a person, or faithfully copy what he
r she does, summons up feelings ranging from
'onder and amazement to downright distaste.

.ndroids and Automatic Devices

'he robots we are most familiar with are the
nes from science fiction that faithfully take on
ıe appearance and actions of men or women.
uch devices, which one might call the most
dvanced forms of robot, should properly be
ılled "androids". The word comes from the
reek words *andros* meaning "man" and *eidos*
ıeaning "form".

Dolls that walk and talk, or do other things to
nitate people, can also be called robots. Many
eople, however, confuse true robots with what
re really simple automatic devices. Gadgets
ıch as clocks or traffic lights work auto-
ıatically, but they are not true robots.

Robots Go to Work

For many years engineers tried to find jobs for robots to do. But, outside amusement halls or the world of films, this proved very difficult. The reason was straightforward. Engineers lacked the skills, and the essential tools, to make robots reliable enough to do jobs in real life.

Two developments changed all this. First, engineers in the United States set their sights lower and decided that robots should do no more than imitate people. They need not necessarily look like people at all. To engineers, robots have simply become mechanical contraptions that imitate the actions of human arms, and are controlled by computers. The second development is that engineering skills, particularly in computers, have progressed at a great pace. The world has become better at making machines that fit the new definition of what a robot is. The result is that, almost overnight, robots have left the kindergarten stage. They have started to go to work.

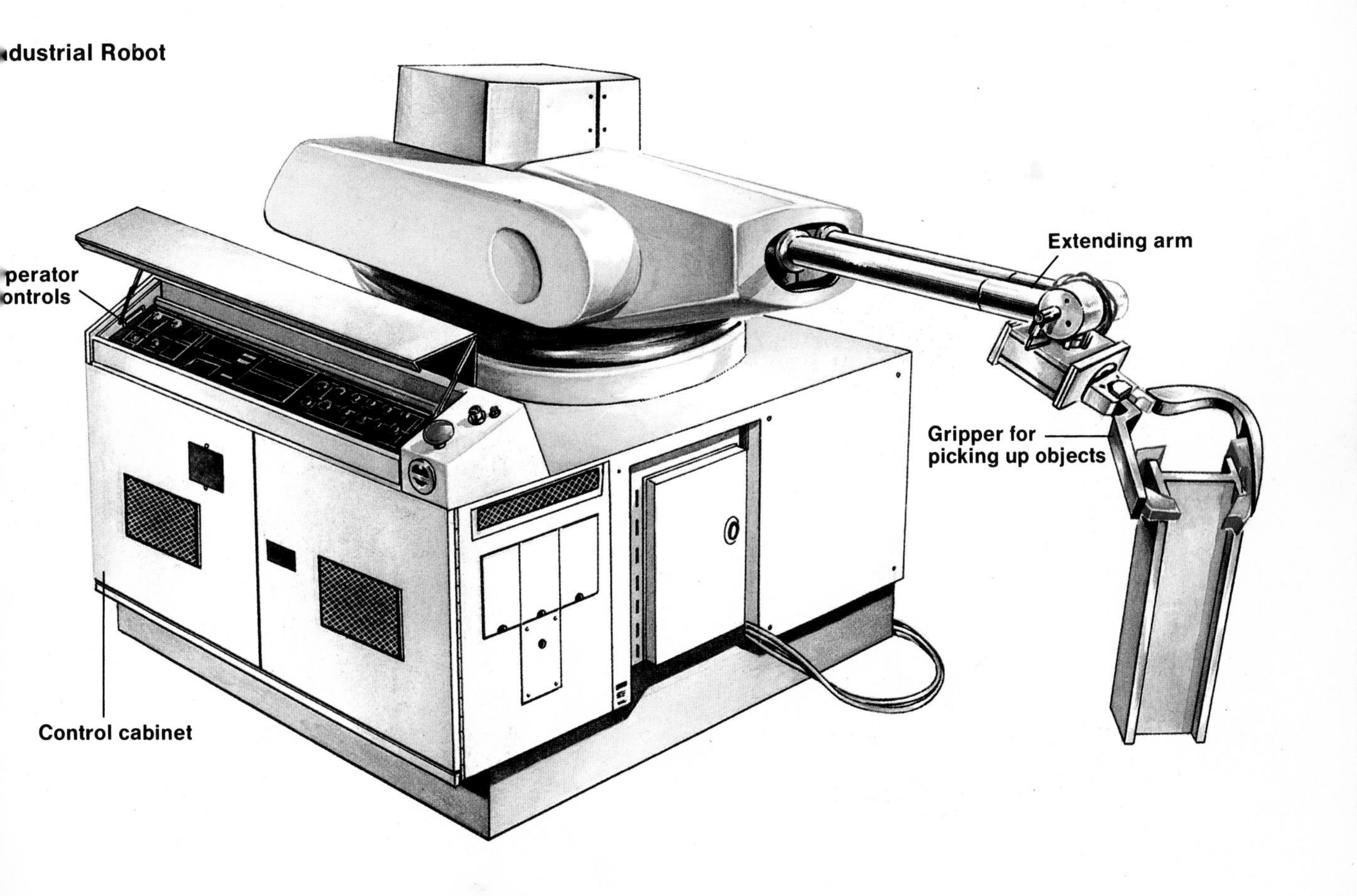

Anatomy of a Robot

The human body contains two sets of components, or parts, connected with action. The first is made up of the brain and nervous system, and the second of limbs and the other parts of the body that move. Robots are made up of components with similar functions. But the average robot at work today is nothing like as complicated as a human being. For one thing, its range of actions is limited by it having only one limb – a single "arm". The instructions that control it, however, are relayed in a similar way to the messages that the brain passes on to the rest of the human body via the nervous system.

In a robot, the job of a "brain" is done by a computer. The computer may store information obtained from the outside world by sensors on the robot. These sensors take on a role similar to eyes and ears in humans. But most robots today are not advanced enough to have sensors. They resemble people who are not only blind and deaf but also lack any way of obtaining information directly from the outside world.

Programming and Electrical Messages

Most robots, then, must receive information in a way that has no direct parallel in humans or other animals. They receive instructions by the mechanism of programming. To program a robot, a human operator feeds the computer messages about the kinds of jobs he or she wants the robot to do.

The really important point about robots is that, by giving a robot's computer a series of different programs, an operator can make the robot do a wide range of jobs. For example, the operator could feed in details about cleaning a window and watch while the robot does it. The next minute, by pressing another button, he or she could tell the robot to pick up a lump of scrap iron. It is this ability to switch quickly from one kind of job to another that makes robots useful in industry.

Like a human being, a robot needs some way to channel the information from its "brain" to its "actuator" – in this case, its one and onl arm. In a human, this channel is made up of series of nerves. Thousands of electrical im pulses stream along these nerves and tell, fo instance, a hand to pick up an apple or a foot t kick a ball. In a robot the information channel made of electrical cables. These drive moto which move the different components in th robot's arm.

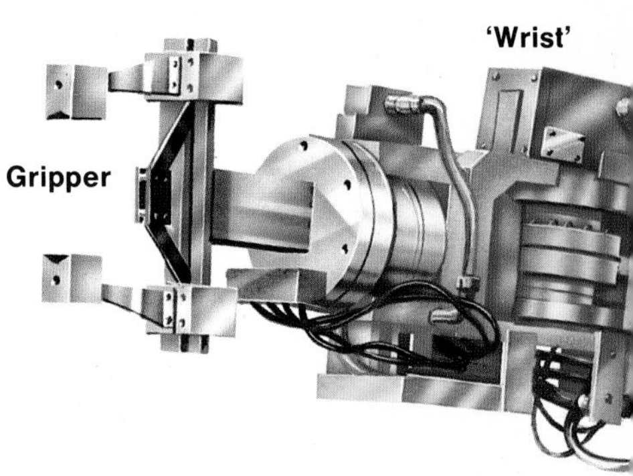

The Robot Arm

Though it has to be content with just one limb, modern robot can do quite a lot with its rathe limited bodily equipment. A robot arm ca position objects to an accuracy of a hundredth c an inch. Some robot machines can lift weights a heavy as one ton. A typical robot arm contains "forearm", an "elbow" and an "upper arm" a well as a "hand". In other words, it is ver similar to a human's. The linkages at th shoulder, elbow and wrist enable the robot ar to move around a total of six axes.

The devices that power the different parts c the arm fall into three types. They rely on th movement of compressed air (pneumati power), on liquids that move under pressur (hydraulic power) or on the power provided by series of electric motors.

One of the most important parts of the robot' body is its "hand". This has final responsibilit for the jobs that the robot does. Robot hand come in different styles for different tasks. Th hand can be a claw, a suction pad, or a magnet. I may also be a special unit into which a tool for particular job may be fitted.

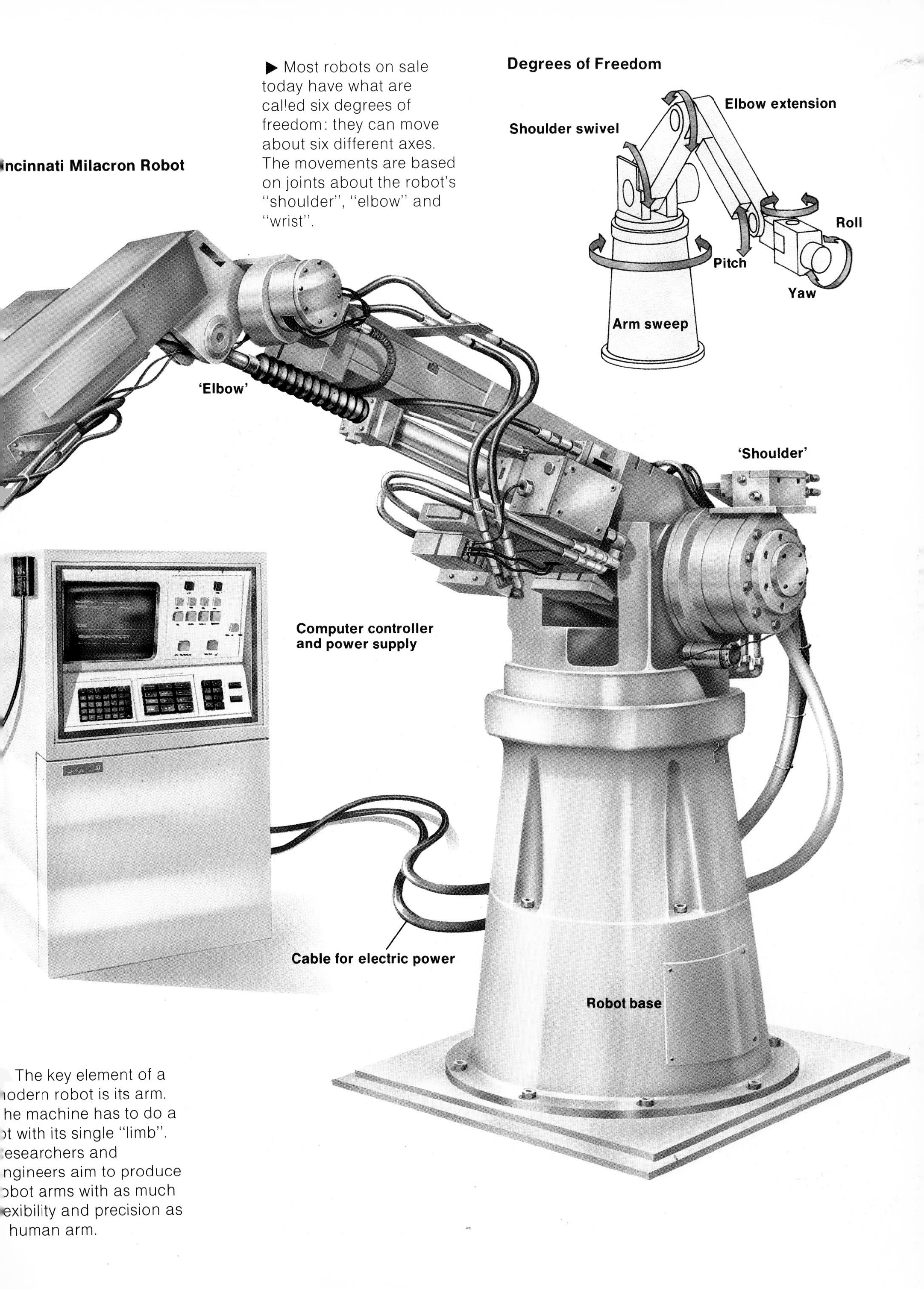

▶ Most robots on sale today have what are called six degrees of freedom: they can move about six different axes. The movements are based on joints about the robot's "shoulder", "elbow" and "wrist".

The key element of a ıodern robot is its arm. he machine has to do a ɔt with its single "limb". esearchers and ngineers aim to produce ɔbot arms with as much exibility and precision as human arm.

Industrial Robots

There are 40,000 or so robots existing throughout the world today. Of these probably 90 per cent work in factories. There is nothing remarkable about this fact. It is just that robots are very good at doing straightforward jobs which involve lifting or some other kind of arm action. The most obvious places where robots can take over these kinds of jobs is in factories.

The Limitations of Factory Machinery

Factories are places in which people make large quantities of goods. Until the Industrial Revolution, which was at its height during the late 18th and early 19th centuries, people made things, but not in a highly organized fashion. Nor did they use machines to do much work for them. Since the Industrial Revolution both the level of organization and the use of machinery have been growing steadily.

So there is nothing new about machines doing the work of people. Until recently, however, most machines in factories did repetitive jobs, or jobs in which skill with the hands and fingers was not involved. For example, an automatic machine might be very good at putting pins through a number of holes in a metal plate. But the pins must always be the same size and shape, and the holes must always be in the same position in the plate. The machine would not be able to vary the job in any way; for instance, by putting a different shape of pin into the holes.

To take another example, machinery in factories has been used for years for channelling liquids through valves and for monitoring their temperature. Such processes are important in many types of industry where the chemical composition of liquids or solids is changed. These industries range from oil refining and steel-making to the processing of foodstuffs.

▼ Here a robot is picking up a casing for a refrigerator and feeding it to a trimming machine.

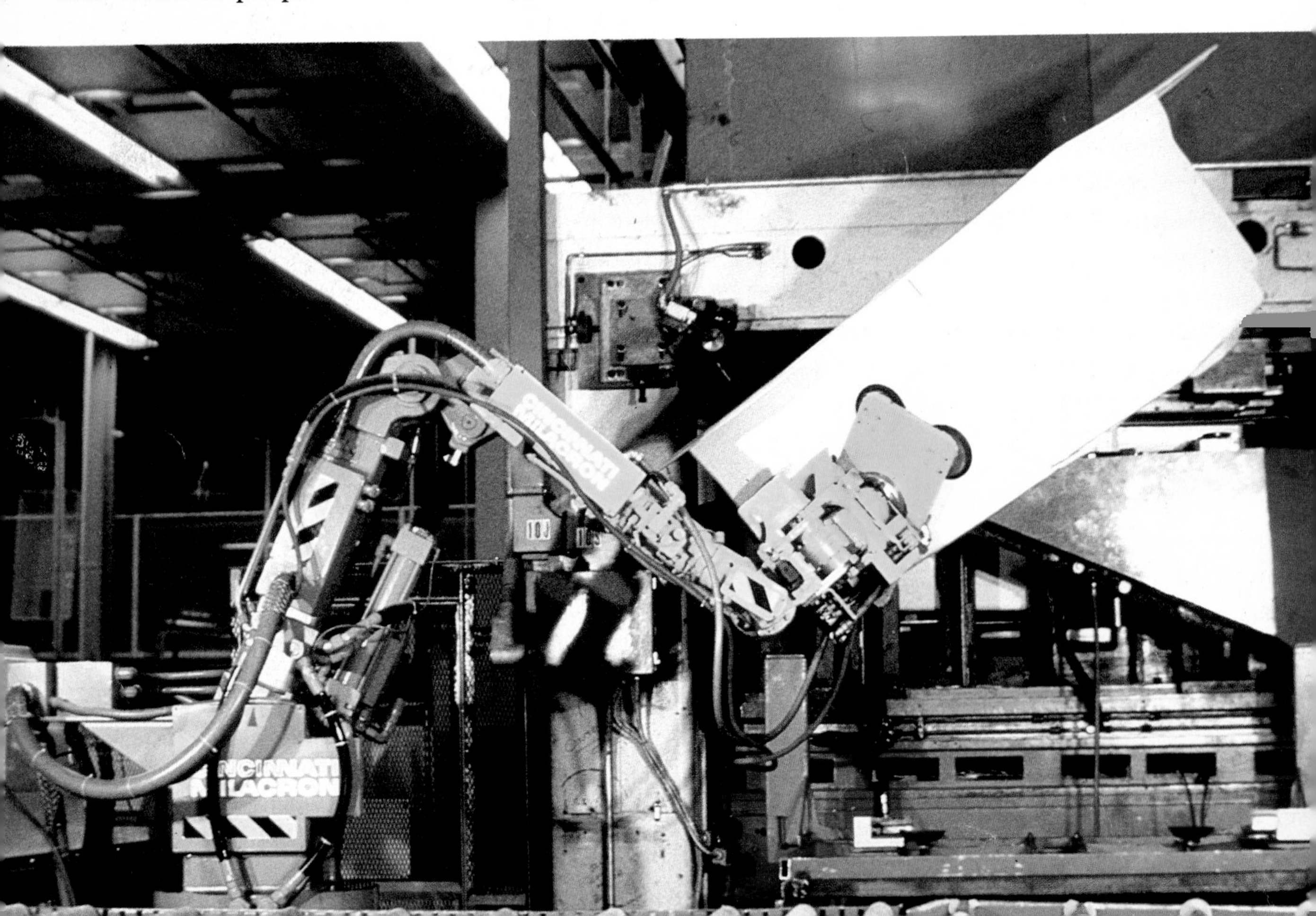

A common job for ıctory robots is spraying aint. The machine's arm an move in a way similar ɔ a person's to follow a omplicated pattern and aint irregularly-shaped bjects. The robot is arefully programmed to o this.

▶ Many industrial products need holes drilled in them. People can do this with tools; but the work is often quicker and more precise if left to a robot. The parts to be drilled are brought to the robots along a moving conveyor belt.

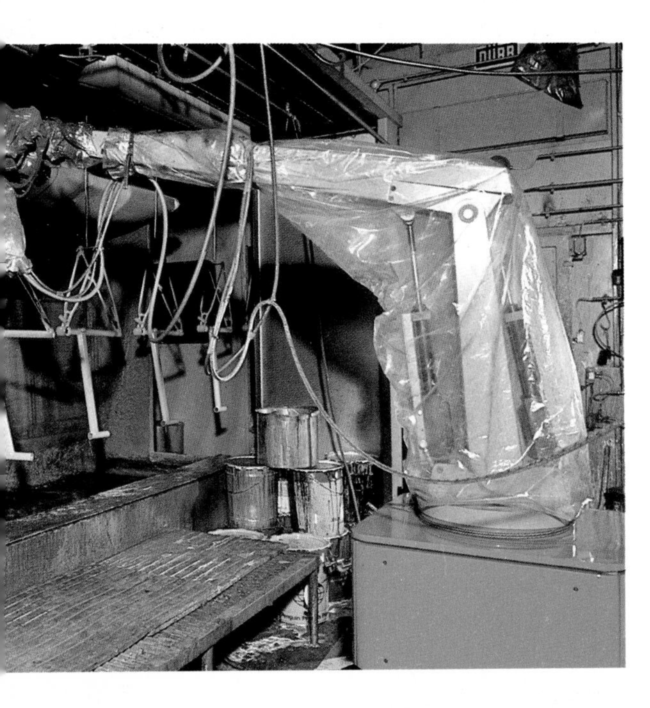

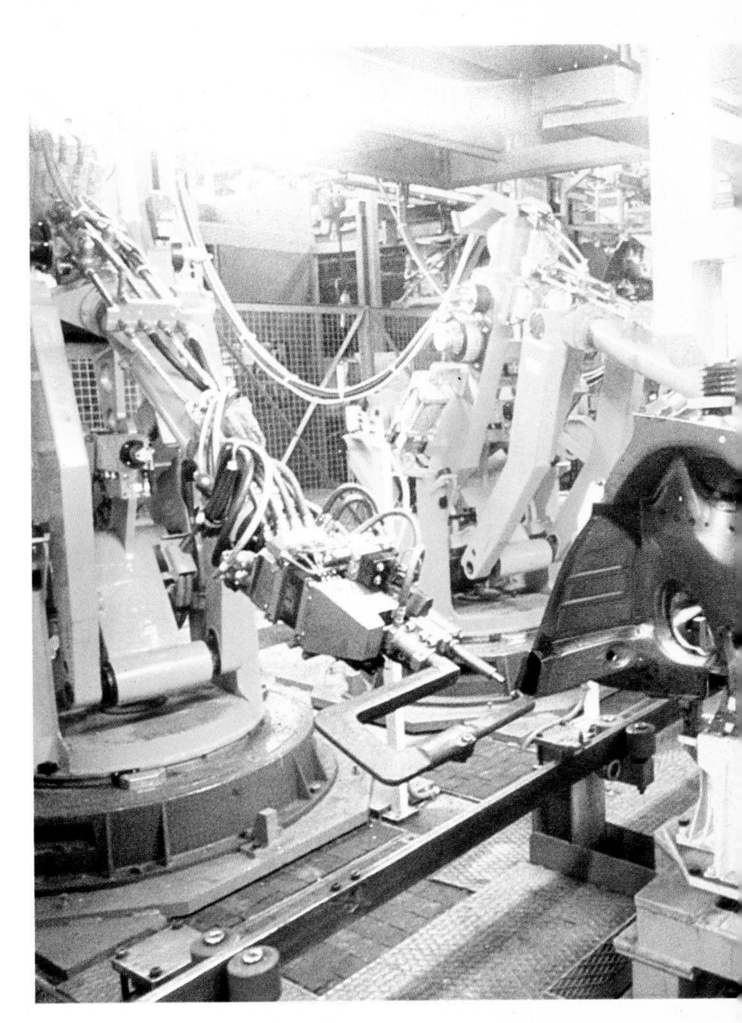

ʒut engineers have found it difficult to build nachines that will, for example, pick up vats of nolten steel and place them in a particular ɔosition. It is in jobs like these, where the work s variable and where a person lifting something ıeeds to be flexible, that robots are now being ısed.

ʈhe Advantages of Robots

ʈwo of the most popular jobs for robots are velding and painting. Robots on car welding ines are a common sight in most parts of the ndustrial world. Robots are also good painters. Armed with a painting gun, they can be ɔrogrammed to coat parts that have complex shapes. They can do this more accurately than a ıuman being can. Spray-painting is often a very ınhealthy occupation for people, as the paint can get everywhere and irritate the eyes, ears ınd nose. For such jobs, robots have advantages ɔver people in terms of health and efficiency.

Another job done by robots is the simple lifting of objects from one place to another in a factory. These can be anything from machine parts to bags of cement. A robot will, for instance, stack the objects neatly according to a set program of instructions. If the factory manager wants them stacked differently, or another set of objects stacked somewhere else, all he or she has to do is change the program.

Robots Working with Other Machines

Robots are also used to fit parts such as electronic components into other pieces of equipment, for example printed circuit boards for electrical equipment. They are good at taking objects out of automatic tools, such as machine tools that cut things, or die-casting machines used in molding. In these jobs, the robot is a servant to another type of machinery and is doing a job which would normally be done by a person.

Robots Diversify

Robots are already starting to move out of factories and into a variety of other places. One job which has always been dangerous for human workers is the inspection and maintenance of nuclear power stations. Robots are now beginning to take over this work.

The central part of a nuclear power station is the reactor core. It is here that a nuclear reaction takes place, generating enormous heat. In a conventional nuclear power station, the heat is converted into electricity. Inside the core are rods of radioactive uranium that provide the fuel for the nuclear sequence.

As the fuel is used up, the uranium rods have to be replaced. Ordinary machines have no trouble pulling the rods in and out of the core, but other jobs have to be done. For instance, the rods must be inspected for faults once they are inside the core. Emergency welds on the interior of the core's wall must sometimes be done.

For some years, mechanisms other than robots have done the essential jobs inside reactors. These are called telechiric devices. Telechiric devices are often confused with robots. They are mechanical arms that respond to the instructions of a human operator. In a nuclear power station, the operator would be sitting out of harm's way in a control room on top of the core. The essential feature of a telechiric mechanism is that a human, rather than a computer, is in control of the arm.

Now, however, a new generation of robots is appearing which is suited for work in nuclear power stations. Taylor Hitec, a company near Manchester, England, is working on a mechanism that can slide through one of the small entrance holes in the core. It then unfolds rather like an umbrella, to reveal a robot hand. A computer above the core, to which the hand is linked by cable, programs the machine to do a particular job. This may be a simple inspection job (when the arm holds a television camera) or removing pieces of metal.

Taylor Hitec is also working on another, more ambitious, project. This is a robot that can take apart the highly radioactive core of a power station when it is no longer required. This is too difficult and dangerous a job for people.

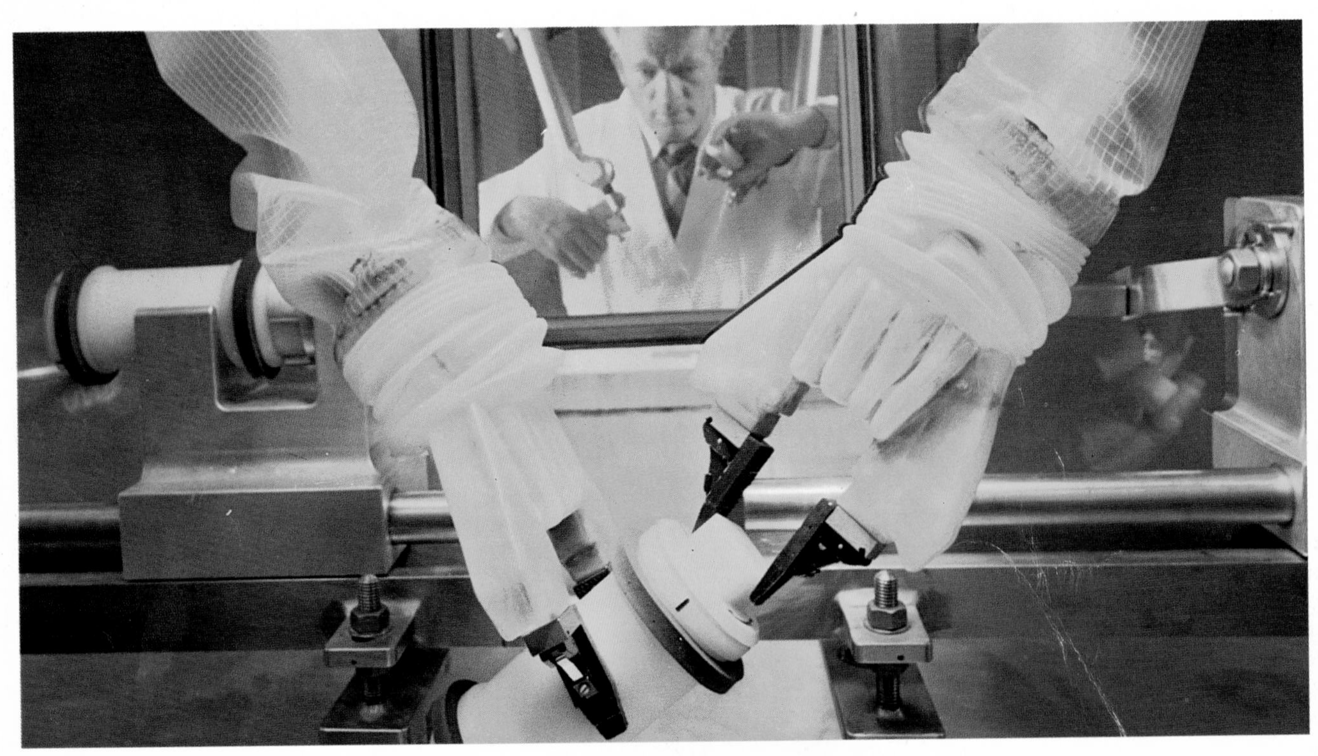

Candy, Underpants and the US Navy

Robots are doing jobs in other unusual areas. Candy firms, for example, use them for putting confectionery into boxes. This is a job normally done by humans but it is boring and can be done more cheaply by a programmable machine.

One large clothing company which makes underpants, wants to put robots to work in its sewing-machine rooms. The robots would grab pieces of material and pass them to sewing machines which would stitch them into the completed garments.

A company called United States Boosters has a design for robots which can shoot high-pressure water to clean up the rocket motors of the Space Shuttle. This would allow the motors to be reused. Also in the United States, a US Navy robot takes out damaged rivets from the wings of aircraft. The US Navy has also designed a programmable mechanism that crawls over the surface of large vessels, cleaning barnacles off them.

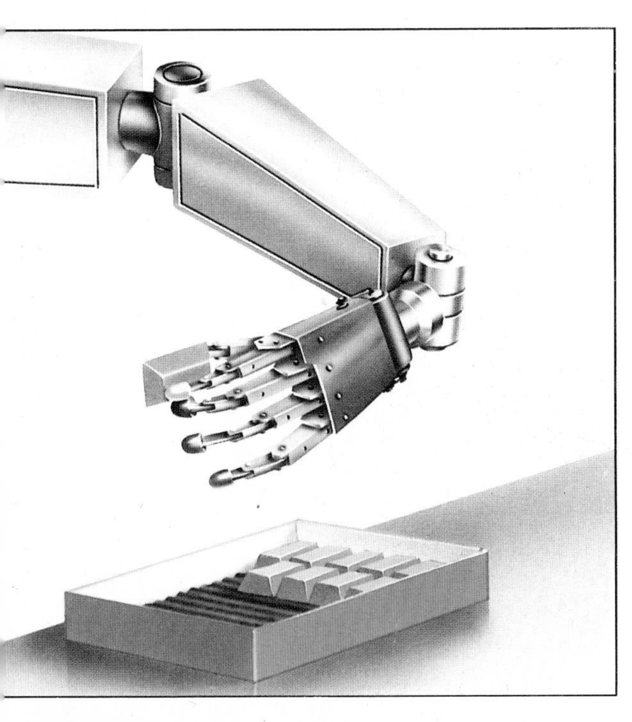

◀ Here a technician is using a telechiric arm to do a complex scientific job. The mechanism merely acts as an extension of the person's arm.

▲ Putting candies in boxes is an irksome, task for people. They can put candies in the wrong position or squeeze them too hard, damaging them. Robots are being called in to take over the job.

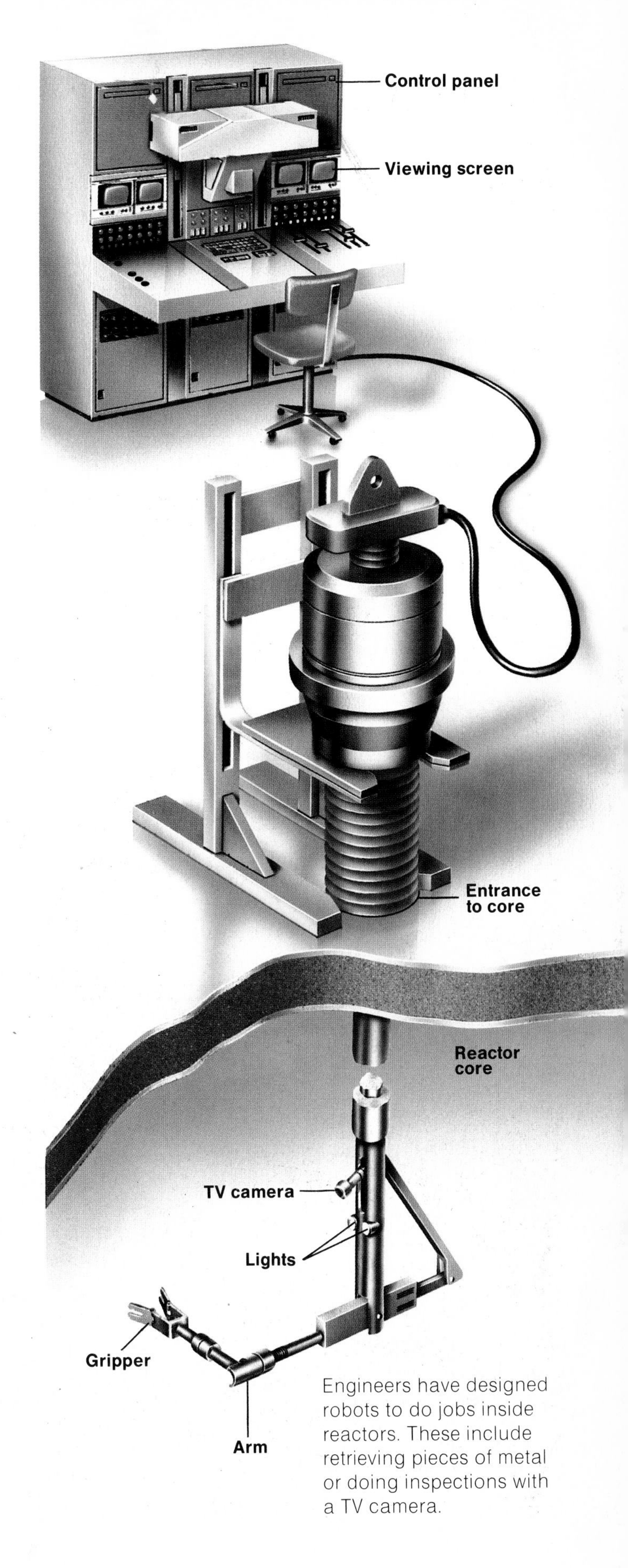

Engineers have designed robots to do jobs inside reactors. These include retrieving pieces of metal or doing inspections with a TV camera.

The Automated Factory

◀ In this automobile factory, car bodies move along a conventional production line and are welded as they pass. A robot welding-line, such as this, is controlled by a central computer. It may produce as many as 1000 car bodies a day.

▶ In the factory of the future, the three main areas of work (planning and administration, design and manufacturing) will be controlled by the same series of computers. There will be far fewer human workers than there are in today's factories. Most of the workers will be needed for changing the programs in the computers.

To understand the impact of robots on our lives, we should look a little more closely at how they link with other, new types of machinery. Robots by themselves, no matter how "clever" they become, will never do all the jobs inside a factory. They are just one part of a mass of computer-controlled machinery that is gradually becoming more important in industry. This is reducing the need for human workers.

The Beginning of Automation

The arrival of the electronic computer in the 1950s brought a new word into our language – automation. This is the name given to the replacement of human effort by machinery that is controlled by computers. Robots, therefore, are one part of the process of automation, but other types of machinery play an important part.

Using Computers and Robots

There are three main areas of work in a typica factory: planning and administration, desig and manufacturing. In planning and adminis tration, the office staff keep a check on th number of goods ordered by customers. The also check the quantity of different parts de livered to the factory by suppliers. Computer play a large part in this work. By using compute terminals, such as display screens and key boards, workers can gain access to a larg computer that contains information about th factory's performance. They can also feed i information that will be useful.

It is a similar story in the design department Draftsmen can work out the shape of nev products on computer terminals. These includ screens on which drawings and other informa

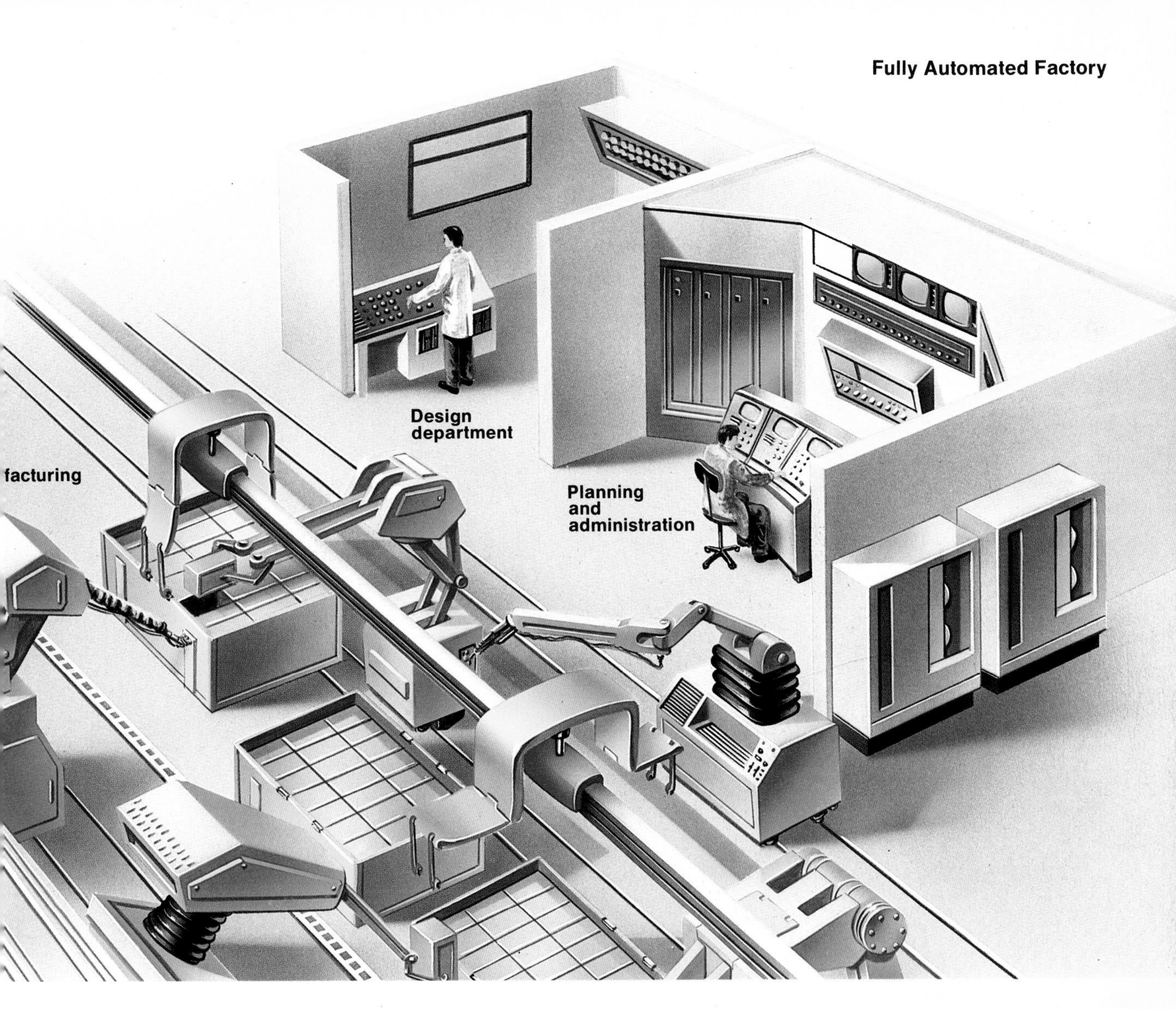

on can be displayed. The draftsmen can also se their terminals to gain access to essential nformation stored in the central computer. This might include details of existing products r the factory's expected orders for the next six nonths.

Robots, as we have already seen, can do many seful jobs in the manufacturing process. They an also work with the machine tools that do utting and shaping operations on the metal, ood or plastic from which the factory's products are made. These machines are also usually ontrolled by computers. Programs instruct the nachines to cut a piece of metal in a certain way, o drill so many holes at particular points in the naterial, and so on. In some advanced factories, ne computer controls a batch of machine tools nd the robots that feed parts into them. The computer may also control other robots that, for example, lift parts from a moving conveyor belt.

Factories of the Future

If these three main areas of factory work can be brought together under the control of the same series of computers, then the factory is about as fully automated as possible. The main feature of such a plant is that few workers are involved in manual operations. Most are in planning and design, plus a few maintenance engineers to keep the machines running smoothly. Most of the workers would be involved in changing the shape and nature of the factory's output by changing the programs in the central computers. Very few factories of this kind exist today, but over the next twenty years they will steadily grow in number.

Teaching Robots

All robots are controlled by computers. But unless a robot has sensors so it can receive information directly from the outside world, someone must put instructions into the computer so that it knows what to tell the robot arm. This is known as programming the robot.

Programming

The most important parts of a computer are the central processing unit and the memory. Instructions in the memory are fed into the central processing unit where they act upon other information (which originally was also stored in the memory). In an ordinary computer, an operator feeds in a set of figures, followed by a set of instructions telling the machine to do a particular operation on the figures. This may be adding up the figures, for example. This set of instructions is called a program.

Computer Codes

The instructions, plus all other informatic inside the computer's memory, are coded as series of electrical impulses, each of which ca be either "on" or "off". This on/off pattern pulses can in turn be represented by a sequen of digits in which 0 stands for "off" and 1 stan for "on". Such a pattern of numbers, in whic only two digits are used, is called a bina system.

Humans, however, would find it far tc complicated to feed information into a con puter using the binary system. Computer er gineers have therefore worked out special cod that automatically translate instructions usir English words into the binary system th computers understand. It is not possible yet use ordinary sentences. Instead, words are fe into computers in the form of special comput 'languages'.

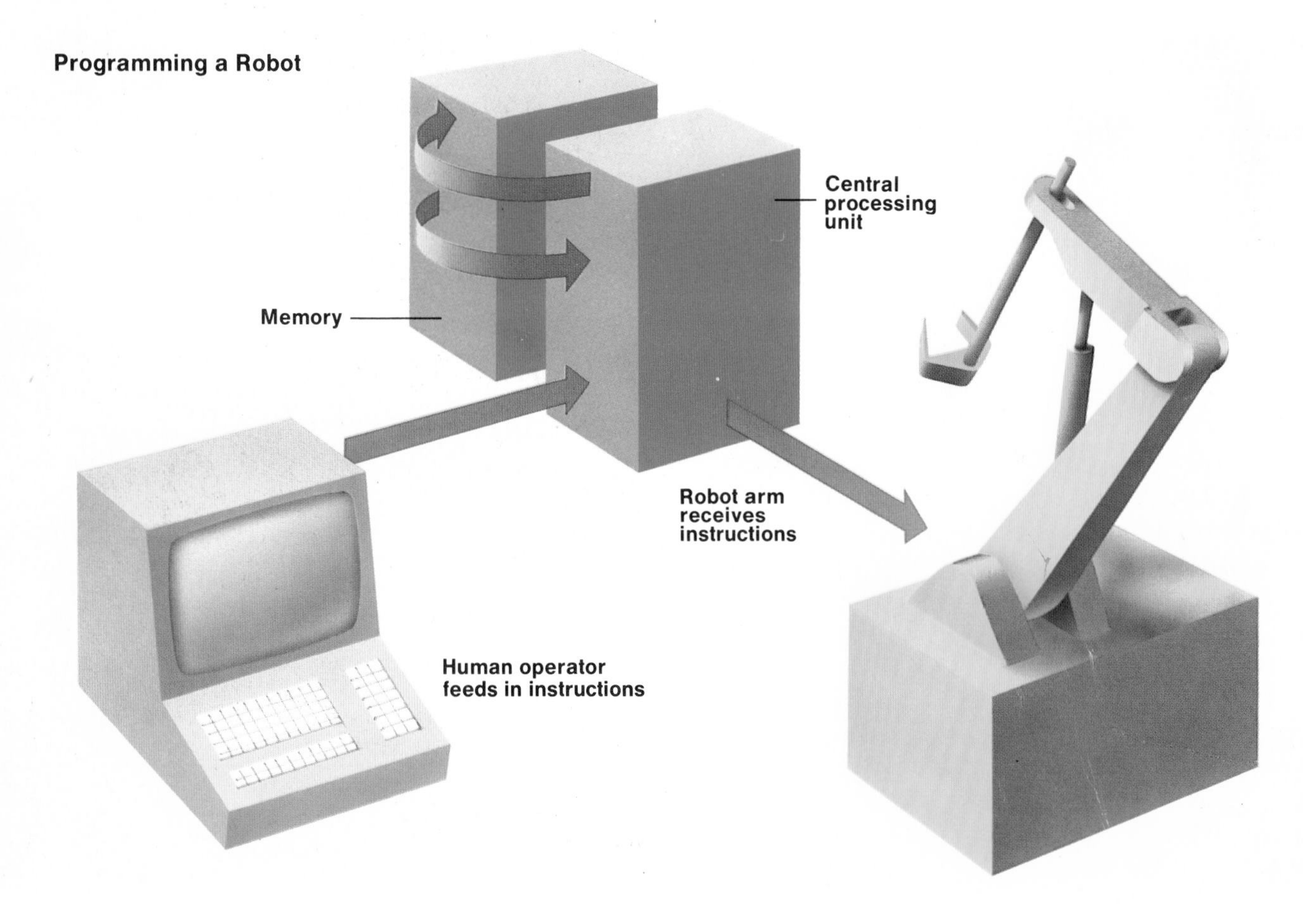

Data

To program a robot, a person normally needs to give it two sets of information, or data. The first relates to the situation in which the robot must work. For instance, the programmer needs to tell the robot how far away it is from a machine tool (if the job is feeding parts into the tool) and details about the shape and characteristics of the tool.

The second set of data concerns the exact job that the robot is required to do. For the above example, this data would include instructions about the number of parts to feed into the machine and the shape of the parts. This set of instructions could also include extra information that the robot could use if something went wrong. For example, if the supply of parts being fed to the robot on a conveyor belt stops for any reason, then the robot could be told to halt operations.

Also included in the program may be the positions that the robot arm is required to take up as it moves through the handling routine. Such information may be necessary to stop the arm bumping into objects while it is doing its job.

Teaching by Doing

A very useful way of instructing robots is called "teaching by doing". In this method a worker guides the robot through a particular sequence of actions. The robot "remembers" the positions by automatically putting the information in its own memory, along with details of what to do at each stage. "Teaching by doing" is particularly useful when the job is normally done by a highly skilled person, and where feeding exact details of the work into the robot would be difficult.

Teaching a Robot to Paint

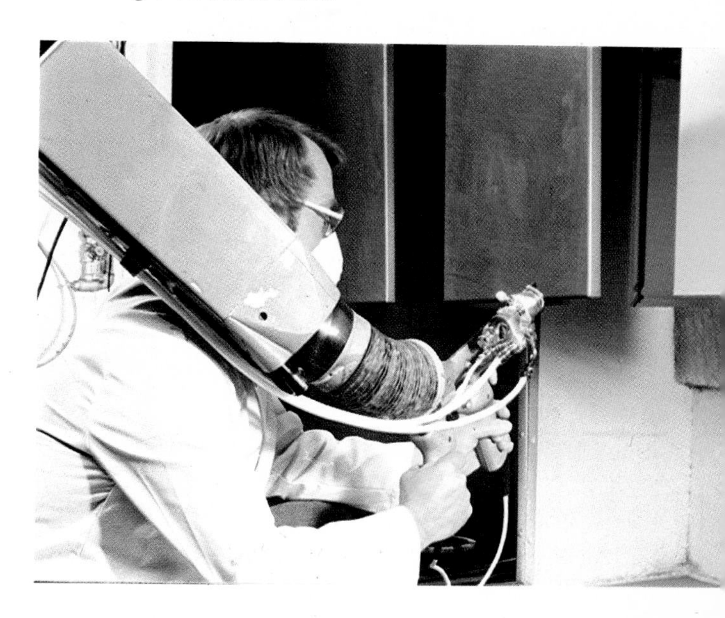

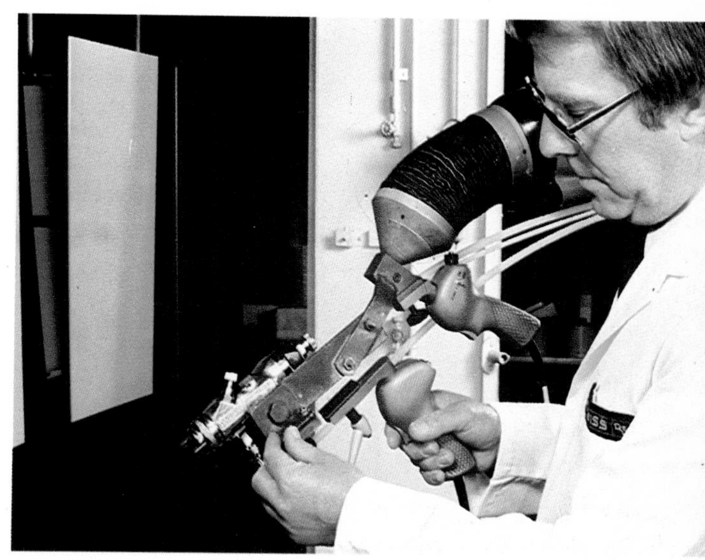

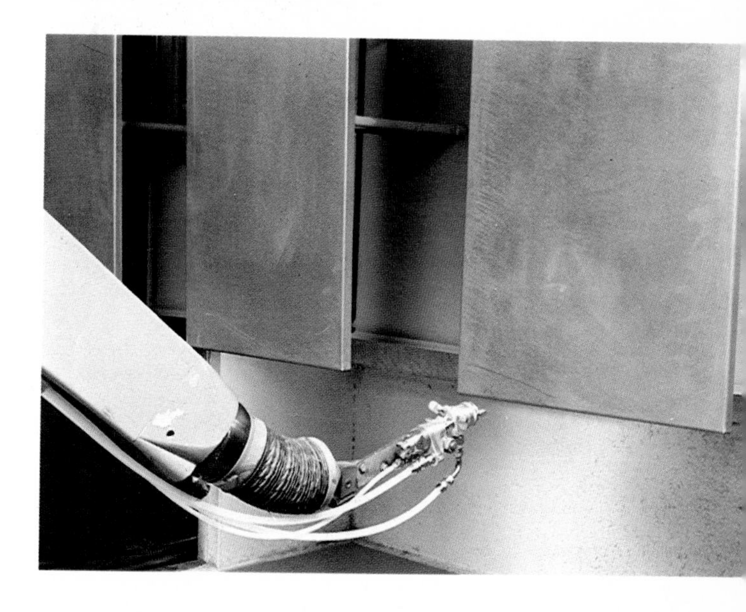

▶ Teaching by doing is the name for instructing robots to do a particular operation by guiding them through it manually. Here (top) a person is moving a robot arm in the same way as he would move his own arm when painting an object with a spray gun. The robot "remembers" the positions for later use. Other instructions (for instance, to tell the machine to change the color of the paint at a particular point) can be given to the robot using a keyboard. Finally (bottom) the robot repeats the movements to do the job as instructed.

Sensitive Robots

All the robots we have looked at so far have been totally dependent on people. Though the robot may do a job without supervision, someone has to tell the robot's computer what to do by feeding in a program. Also, if something happens during the job that the robot has not been programmed to expect, it is powerless to react.

There is, however, a way around this problem. This is to provide the robot with the kind of mechanisms that people have for receiving information from the outside world and channeling it into the brain. Engineers call these mechanisms "sensors". The most important human sensors are the eyes, ears, nose and nerve endings.

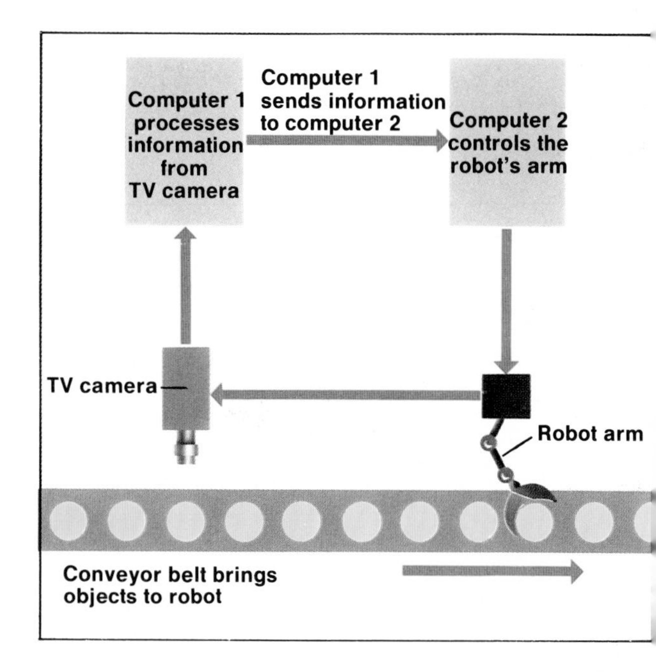

▲ A "seeing" robot needs two computers. One computer controls its arm and the other computer makes sense of information provided by a TV camera.

Robots that "See" and "Feel"

The most usual way of giving robots sense is through vision. A television camera takes pictures (perhaps of an object that the robot is supposed to pick up) and feeds the information to a computer. This computer is normally separate from the one that controls the robot's

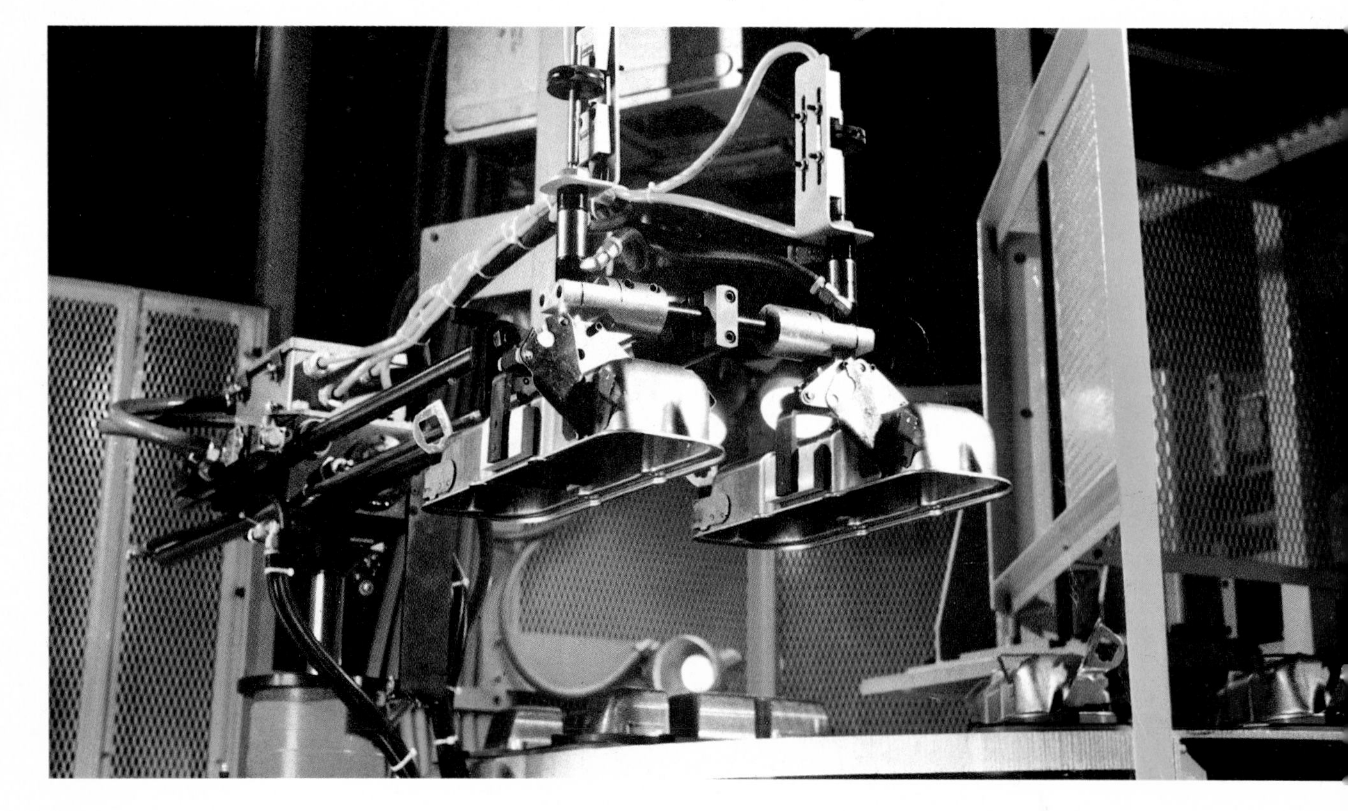

▼ Few robots with vision operate in the world's factories. Here one of them goes through its paces, "seeing" objects with a TV camera so it can pick them up.

▼ To judge whether a room is dark enough to switch on the light, a robot needs a light meter controlled by a computer, instead of eyes and a brain.

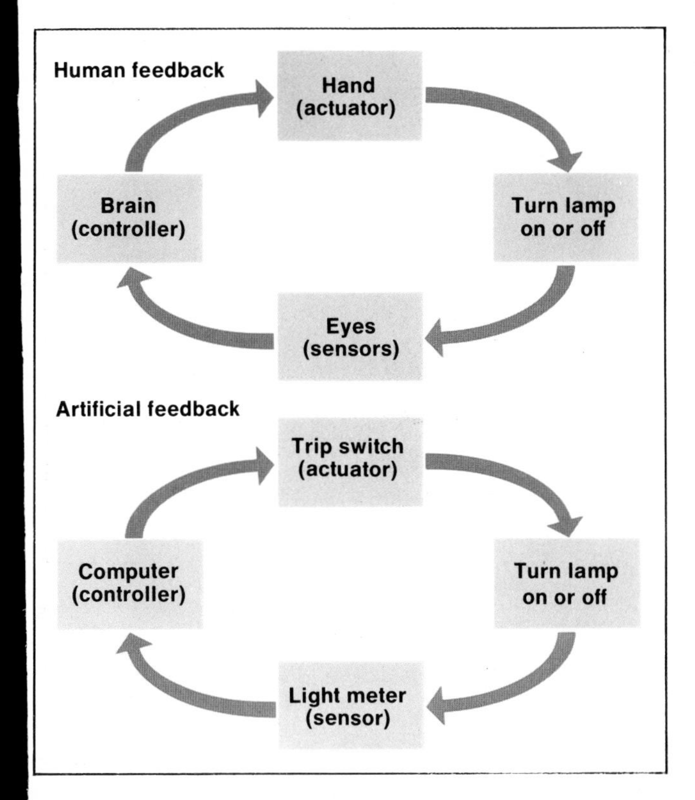

arm. The computer has been programmed to deal with, or process, the information supplied by the television camera. It changes it into a form which can be used by the computer controlling the robot arm, where it is added to other information already stored there. The robot now has enough information to carry out its job: instructions on what to do and a picture of the objects and tools it is working with.

One obvious advantage of a robot with vision is that it can be fed a series of differently shaped parts and handle each of them perfectly well. It will recognize each one and its "hand" will know the best place to pick it up. A "blind" robot would try to pick each one up in the same way, possibly with disastrous results.

Another way in which robots can have sense is through touch. Engineers can fix force-sensors to the end of a robot's hand. When these sensors touch something, information about what is happening is sent back to the robot's computer.

▶ To weld something accurately, a sense of touch is needed. Here a welding robot senses the position of a piece of metal with pressure sensors.

For example, when the robot bangs into a wall the force-sensors send that information to the robot "brain" which then tells the robot to change its actions so it doesn't bang into the wall again.

Feedback

The examples of vision and touch-sensitive robots demonstrate the important idea of "feedback". "Feedback" is the way in which the actions of people or things may be altered or changed by outside events. People use feedback all the time in their day-to-day lives. It comes into play, for instance, when you walk down a road, see a tree in your path and step to one side to avoid colliding with it and injuring yourself. Attempts to give robots feedback are necessary if they are to become useful in several different situations. Unfortunately for engineers, however, equipping robots with this ability is not easy. It is very difficult to code information from the outside world in a way that makes sense to the robot's computer, and then to make sure it processes the data quickly enough for the robot to respond. At present, probably only five per cent of the world's robots have sense. They are known as second-generation robots.

Robots on the Move

Armed with sensors, robots could have a great many uses in areas outside general industry. A popular place for them might be in the home. A Japanese company is spending about $300,000 a year on research to make home robots feasible. Such robots could, for instance, open the door when you arrive home, bring you your slippers and then pour you a drink.

Robots on Wheels

Robots would have to be mobile, as well as having sensors, to be able to find their way around. In Britain, engineering groups at GEC in Rugby and at the University of Warwick are trying to build robots on wheels. These machines are designed for jobs in factories, but they could be just as useful in the home. They would need a source of movement such as an electric motor, and a feedback system to send information about the outside world back to a steering mechanism.

Mobile robots would have a lot of other uses. For instance, they could work on building sites and down coal mines. For many years, engineers have dreamed of putting robots to work in unpleasant places such as mines. Sensors would guide them along tunnels to the coalface. At the coalface, the robots would cut the coal and hoist it on to trucks.

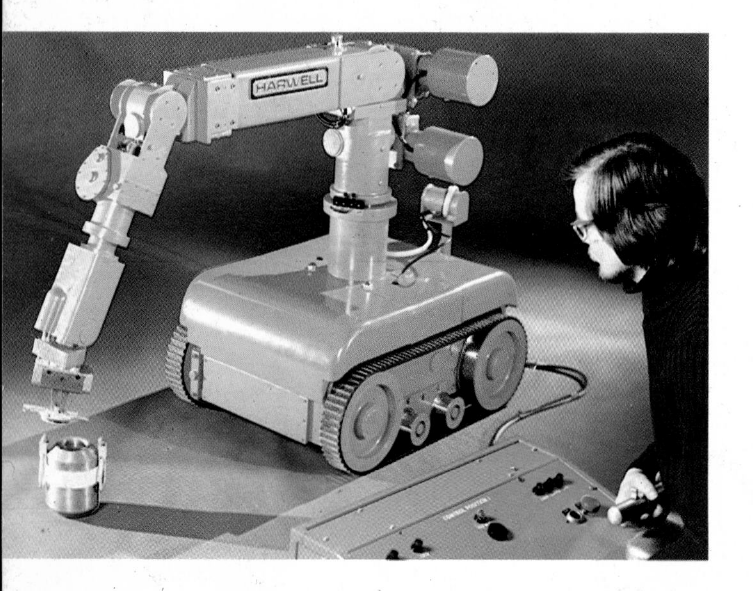

◀ Robots that can move have been the goal of engineers for years. In an early design, the UK Atomic Energy Authority devised a rudimentary robot, shown here, that moved around on caterpillar treads.

Robots on the Seabed

A still more promising place for robots could be under the sea. The bottom of the Pacific and other deep oceans is littered with potato-shaped lumps of rock that contain rare metals such as manganese and nickel. In a world that is gradually running out of its most precious metals, engineers have been concerned with finding a way to bring these lumps to the surface. Conventional methods would drag a series of sleds along the bottom of the sea or suck up the rock with a machine resembling a giant vacuum cleaner. But robots could actually crawl over the seabed looking for the most promising lumps. They would then load them into buckets which would carry them to the surface. The robots would have to find their way around seabed objects. They might even have to swim and take avoiding action if pursued by sharks. A coat of rustproof paint would also be necessary.

"Insect" Robots

Most mobile robots use wheels to get around, but some engineers are studying the possibilities of robots on legs. Although they look more like people, two-legged robots are frowned upon in engineering circles. Research workers have found it difficult to make robots of this kind move around easily. They are ungainly and keep falling down. Robots with six or even eight legs are much more practical. There have even been hints that robots of this kind are being developed in the Soviet Union for use in a possible war. Such "insect" robots would be very good at traveling over rough ground where ordinary vehicles, such as tanks, are unsuitable.

▶ Mobile robots of the future may operate under the sea (top) and down mines (bottom). The "swimming" robots would pick up metallic nodules from the sea-bed, or perhaps bring shipwrecks to the surface.

Robots in Space

Scientists have already made great use of machines in space. Although many people would love to copy astronauts and take a trip into the heavens, the truth is that space is not a very comfortable place. It is cold and dark and lacks all the comforts that make life pleasant, or at the very least bearable, on Earth. Still more important, the air we need to breathe to stay alive is missing in space.

Machines have enabled people to make use of space. Communications companies have put satellites into orbit. These send information from one part of the Earth to another. Other spacecraft send us information about the weather and all sorts of scientific data. For example, they send us information about the nature of the Solar System, and pictures of other planets.

Viking and *Voyager*

Most of this space hardware is definitely not i the robot category. It ranges from automati machinery that regularly sends scientific data t Earth, to equipment which connects telephon calls without a human intervening. A fe spacecraft, however, have come near to bein robots. These include the *Viking* landers th the United States sent to Mars in 1976. The had scoop-like devices that dug into the soil t take samples. These were controlled by com puters on board the landing craft. The late *Voyager* spacecraft, which the United State sent deep into the Solar System in 1977, contai computers. These can guide the craft in particular direction. The computers can b programmed by radio waves sent from Earth.

pace-farer Robots

ar more familiar robots may appear in space ver the next few years. They would be more dvanced than the robots that already toil away ı factories on Earth. One exciting possibility ›r robots is in tending space factories of the ind that may become established around the ıd of this century.

The different characteristics of space (low ravity, no air, and abundant energy from the un) make it possible that some industrial rocesses could take place there under far more ıvorable conditions than on Earth. People 'ould probably not want to work in factories ut in space because of the unpleasant living onditions. It would be like working on an ffshore oil rig, only far worse and much farther om home. These problems have led to the idea f recruiting a new breed of robot space-farers.

These robots might be fitted with little rocket ıotors so they could whizz their way around. 'hey would also need suction grippers to walk around in a low-gravity environment instead of falling off into space. As they become more advanced and more skillful, the space-robots could work on other jobs – mending faulty spacecraft, for instance. They could also build huge platforms to house large panels of solar cells that would capture the Sun's energy and send it to Earth as microwaves. This would make a big contribution to solving the world's energy problems. Mobile robots on wheels could also trundle over the surface of planets, acting out the part of explorers.

Putting robots to work in space, instead of people, will also save money. According to calculations made recently, keeping a person in space for one hour and making sure he or she comes back to Earth alive costs about $15,000. Space robots would be expensive to develop, but once in orbit they would cost far less than humans to keep in operation. Furthermore, if a space-robot should break down, it could easily be replaced.

◀ The American space robe *Voyager* has begun tour of the outer Solar ystem. A high point will e when it approaches eptune in 1989. The robe is a prototype of the ompletely independent pace robots we may see the future.

▲ American space engineers have designed the Mars Rover, a robotic vehicle that could trundle around Mars. The vehicle would have sensors, such as TV cameras, to pick up information from the surroundings.

▶ The USA landed two *Viking* probes on Mars in 1976. They were able to analyse soil samples and the Martian atmosphere.

Robots in Fiction

◀ The robot Gort, a 12-foot giant, appeared in *The Day the Earth Stood Still* (1951).

▼ Tobor from *Tobor the Great* (1954), was a robot hero who took care of his inventor's grandson.

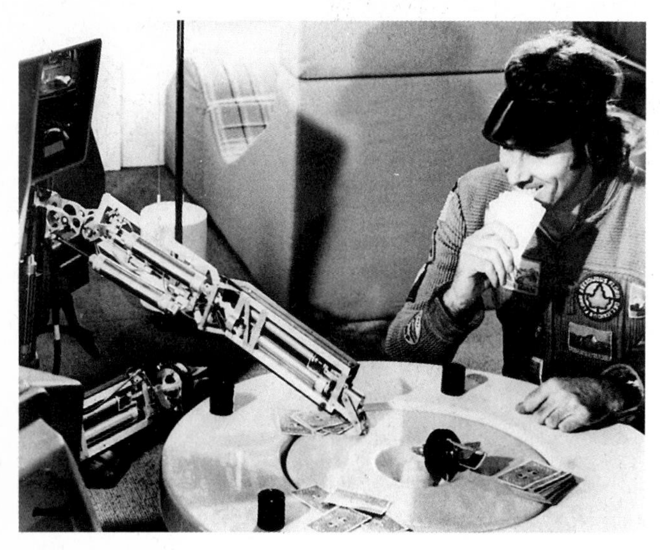

◀ In a film called *Silent Running* (1972), a robot demonstrates its ability to play cards with a human.

▶ The film *Star Wars* (1977) showed robots as friendly, lovable creatures. Shown here are C-3PO and R2-D2.

The Meaning of "Robot"

The word "robot" has its roots in fiction. The term came into the English language from the Czech word *robota* meaning "servitude". It is widely used mainly as a result of a play, by the Czech writer Karek Capek, called *Rossum's Universal Robots*, or *R.U.R.* The play was written in 1920 and translated into English three years later. It was first performed at the National Theater in Prague in 1921.

Rossum's Universal Robots

It is worth knowing something about *R.U.R.* because it set the tone for future plays and films about robots. In *R.U.R.* a factory, run by a man called Rossum, makes android machines that actually look and behave like people. The androids have a great capacity for work. At fir Rossum sells plenty of them to other factorie But after a while events take a sinister turn. Th robots learn how to think for themselves. The rebel against their human controllers an gradually take over the world.

This kind of turn-around appears in much c the fiction that followed *R.U.R.*, and man stories portrayed robots as evil and threatenin Even the words that Capek gave to Radius, th chief robot, are similar to those that followin generations of robots have spoken as the dominate the cowering humans. By gainin possession of the factory, Radius explains in on of his speeches, "We have become masters c everything. The power of man has fallen. A ne world has risen. The rule of the Robot."

◀ The Daleks, from the TV series *Dr Who*, are metal-clad visitors from outer space who wage war against humans.

▼ Robots have also featured in exhibitions. Here one called ON/OFF, and a companion, go through their paces.

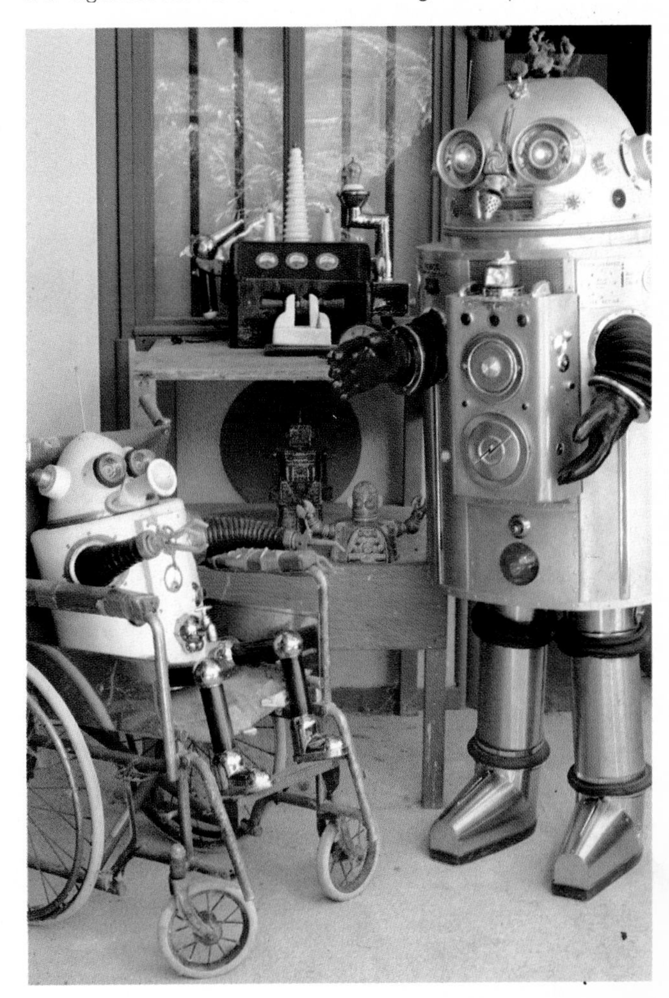

Robots in Films

Robots, mostly androids, also feature widely in films and on television. In *Target Earth*, a film made in 1944, the story is familiar. Alien robots from another planet attempt to take over the world. *Zombies of the Stratosphere* (1935) and *Phantom Empire* (1952) are both based on adventures in space. The robots in these films are also threatening. It was left to the film *Star Wars* (1977) to portray robots as friendly, even lovable, creatures that actually help people rather than try to kill or suppress them.

The "Three Laws of Robotics"

The writer Isaac Asimov has played a major part in presenting robots as beings with whom humans can find some kind of fellow feeling. In some of his stories the robots are made by a company called Robot and Mechanical Men Inc. This company employs a robot psychologist to make sure that each machine's brain works in such a way that the robot will not cause problems to people. Asimov even went to the trouble in 1940 of devising his famous 'three laws of robotics'. These are a sort of code of conduct designed to ensure that robots will act for the good of men and women.

The laws are:

1. A robot must not harm a human being, nor through inaction allow one to come to harm.
2. A robot must always obey human beings, unless that is in conflict with the first law.
3. A robot must protect itself from harm, unless this conflicts with laws 1 and 2.

Intelligent Robots

None of the robots we have looked at so far matches up to humans in one important respect – thinking. Even "second-generation" robots, the ones with sensors picking up information about the outside world, come a long way behind people in their ability to make sense of this information and act accordingly. "Second-generation" machines react simply as a reflex to what is taking place around them. The ability to weigh up the situation, perhaps comparing it with previous experience, is beyond them.

Third-generation Robots

But how do engineers propose to make the ne style of computer? Two ideas are at the center o this work. First, researchers would like t change the way computers process their instructions or programs. At present, nearly all th computers in the world handle instructions on after another, or in sequence. This is becaus computers' electronic units are not very complex. Their layout is very simple when compared with the human brain.

◀ A robot with "intelligence" is being designed by the US Navy. It will be able to make some decisions on the basis of information that i acquires from its surroundings. The machine will be able to swim under water and repair pipelines.

Artificial Intelligence

"Thought", then, involves a rather more complex method of processing data than is possible with the computer "brains" of today's robots. The answer is to improve these computers. This brings us to an important area of research called artificial intelligence, or AI.

With AI techniques, computers could become far more powerful and versatile than they are today. They would be able to make decisions and so could be excellent substitutes for human workers. The new, AI-assisted computers would also act as the brains for a new breed of third-generation "intelligent" robots. Such machines would resemble the androids of science fiction because they would actually think like people.

In the human brain many lines of thought, o processing of ideas, can take place at once – i parallel. This is why humans can think of man unrelated ideas at once. Some people are able t connect these ideas to arrive at startling conclusions or to produce brilliant pieces of work. Thi is how, for instance, Sir Isaac Newton was abl to devise his laws of gravity and Bach t compose his music.

Designing machines with this element o parallel processing is not easy. Different strand of "thought" patterns must be set in motion an then linked up with each other again from tim to time. Making a machine that works like this i rather like planning a complicated railroa network. Such a network must make a larg number of connections and every train must b n time – down to the nearest split second.

The second way to make intelligent computers is by redesigning the memories. Where today's computers have one memory, the new machines would have two. One would store programs and data relevant to the job in hand and the other would be a "knowledge base". It would store the great mass of general knowledge that makes it possible for humans to move around and function normally.

Linking up the "knowledge base" with the computer memory and the machine's central processing unit is necessary if "third-generation" robots are to become reality. Researchers are already working on these problems.

Factory robots could be of greater use if they were mobile. For instance, robots of this kind could trundle around machine-shop floors repairing tools or delivering goods. Engineers are developing such devices. These could be on sale in the late 1980s.

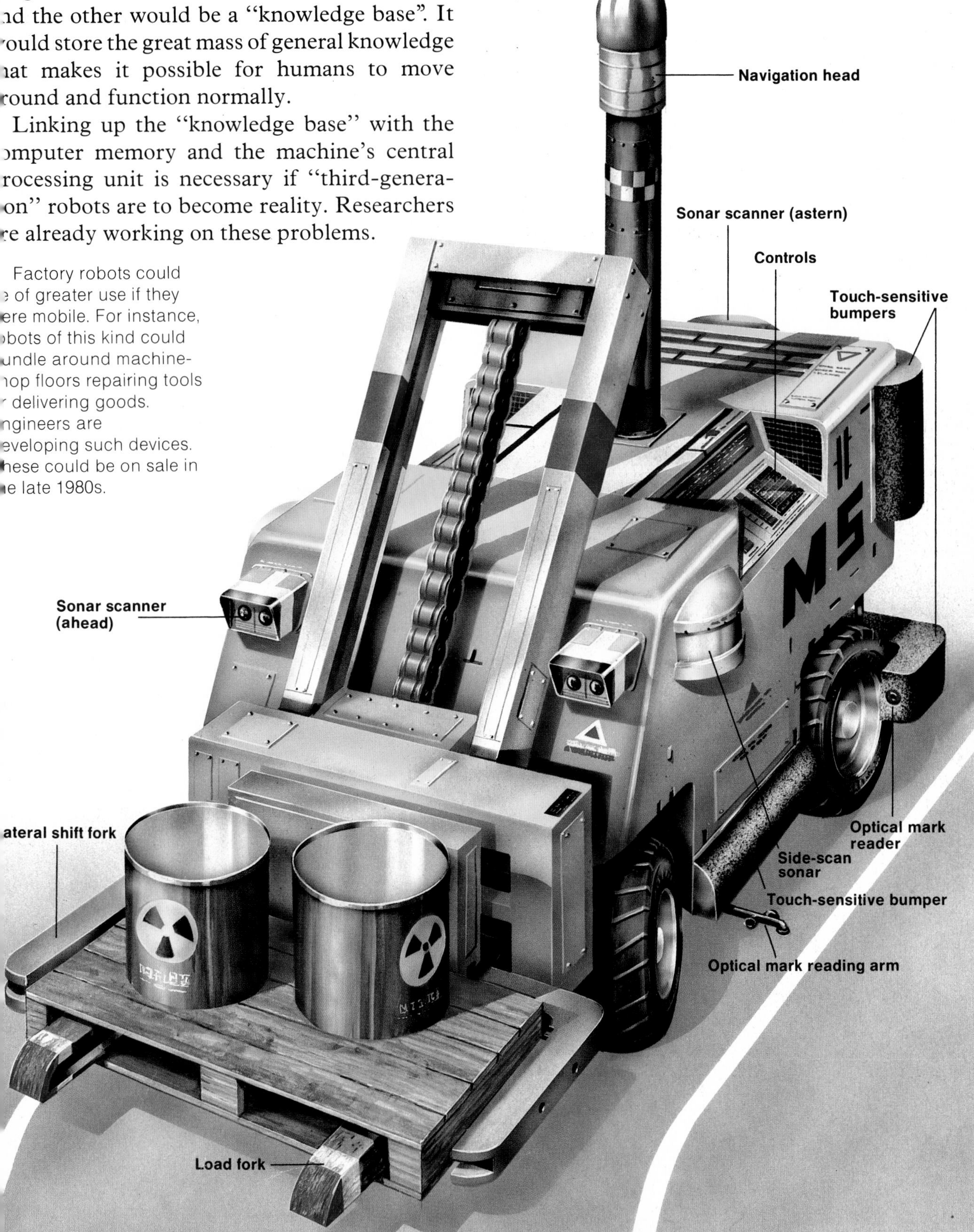

Robots Built by Robots

In the most terrifying science fiction stories involving robots, they first take over the world and enslave the human population. Then, and this is the point at which shivers start to crawl down people's spines, the robots start to breed. Although robots that can reproduce themselves have not yet appeared in real life, researchers in the United States have recently laid down the engineering guidelines which they would probably follow.

John von Neumann, a Hungarian-American scientist who also did much of the early work on computers, started it all in the 1940s. He proposed that machines should be able to reproduce by following a given set of rules. Scientists later found out that these rules are remarkably similar to the way that animal cel reproduce themselves.

According to von Neumann's theory, a sel reproducing robot system would need fou parts. The first is an automatic factory whic collects raw materials and turns them int products according to written instructions. Th second part is a duplicator which takes a writte instruction and copies it. The third part is controller which sends the instruction to th duplicator for copying. One copy is passed t the factory for action and the other to the fin parts produced by the factory. The final part the system is the written instruction itself whic tells the factory to construct a complete ne automatic factory out of the parts it has made.

actories on the Moon

cientists played around with these ideas for ears but then, in the summer of 1980, re-earchers employed by the National Aero-autics and Space Administration (NASA) roduced an engineering blueprint of how a elf-replicating system would operate. The way works is shown in the illustration.

The system has four main parts. There is a rocessing unit which mines materials and efines them. There is also a production system hich makes parts from what the processing nit provides. Another production system turns ut complete products from individual com-onents. Finally there is a "universal con-ructor". This constructor is the key to the hole system. It takes parts from the produc-on systems and builds complete new factories ith them. As the new factories contain their wn universal constructors, the whole process can begin all over again. In this way the factories can reproduce themselves at a fast rate.

In each universal constructor is a series of robots. These know how to make parts for the factories themselves, but they are also intelli-gent enough to turn out new robots. An interesting point about the NASA exercise is that it is hoped to put the system to work on the Moon one day. Such a space factory would make great use of the valuable raw materials on the Moon, and it would never need to be serviced from Earth. It could continually provide itself with a new supply of robot workers.

▼ People will probably return to the Moon in the 1990s or early next century to mine raw materials. The moon factories will quite likely be run by robots. The robots may self-replicate so that the factories have an expanding workforce

The Robot Future

Since factories were first built, people have been eager to invent new kinds of machinery that will do the jobs of factory workers. As industrialization increased during the 19th century, factory owners found they could produce more goods, not by taking on more workers, but by installing new machinery.

At the beginning of the 18th century in Britain, about 80 per cent of the population of 5 million worked on the land. By the middle of the 19th century half the country's working population had jobs in factories. From the beginning of the 20th century this proportion has steadily fallen to about 35 per cent today. Other industrialized nations throughout the world have followed similar patterns.

Robots – A Mixed Blessing

The widespread use of automated machinery, computers and robots will continue this trend. Fewer and fewer jobs in manufacturing industries will be available to people wanting work. Many will find jobs in other industries that are growing, such as the "service" industries – catering, local government, banking and transport. But just what will happen is difficult to tell.

Employment in the Future

In the past, new types of employment appeared for workers whose jobs had disappeared. In the 19th century, for example, millions of jobs were created by the growth of factories. Yet, only a few years before, no one would even have been able to tell what went on in such strange places.

On the other hand, robots taking over jobs in factories could worsen the high unemployment rate that the industrialized world already suffers from. Although new industries (such as making and programming electronic goods) are being created, they require few workers. Many people are worried that not enough jobs will be created to absorb the people no longer needed in factories.

Learning to Live with Robots

The development of robots will have anothe perhaps deeper, effect on society. Robots wi make their presence felt not just in factories bu in many other areas – in the home, for exampl and in hospitals and other service jobs. Peop have had little trouble in getting used to ne forms of electronic gadgetry such as washing machines, home computers and video reco ders. But they may find the new forms of robo more difficult to accept easily.

The "intelligent" robots discussed earlier ma produce mixed feelings among people who wor with or control them. Some writers argue th people will welcome machines they can get o with easily. There would be no need to addre the new breed of robot with special compute languages. Most probably people would com mand them simply by talking to them. But, o the other hand, men and women may feel uneas when confronted by machines that are a "clever" as, or cleverer than, themselves.

The Robots of the Future

One thing is sure. The robots that wande around homes and factories during the ne century will not be the metal-clad, cumbersom monsters of past science fiction films. They wi be specially designed to fit in with the surroundings. The robots of the future wi probably be sleek and smooth, and look lik ultra-efficient vacuum cleaners. Robots wi gradually become as familiar as the electri motors that abound in homes and factori today.

▶ Autonomous machines will do many of the world's mundane jobs a century or two from now. It is impossible to predict how far into everyday life the machines will intrude. But undoubtedly the jobs that people find the most unpleasant will qualify firs for the robot takeover. We may well see robot "life savers" rescuing people from floods, as shown here. Robot fire engines and robot lifeguards on the beach are further possibilities.

Glossary

Actuator A device responsible for action, e.g. robot hand, car wheel, person's foot.

Android A machine that both looks and behaves like a human being. The word 'android' comes from the Greek words *andros* ('man') and *eidos* ('form').

Artificial intelligence (AI) The discipline by which engineers try to make computers copy human thought-patterns. AI-assisted computers will be used as the 'brains' for future third-generation robots.

Automatic machine Equipment that operates according to a fixed set of instructions. Not to be confused with robots. Automatic machines include devices such as clocks and traffic lights.

Automation The process of linking up computerized tools and machines and setting them to work with the minimum of human intervention.

Binary system The code of 0s and 1s which makes up a computer's "language".

Central processing unit The part of a computer that processes instructions from a program.

Cincinnati Milacron A popular form of industrial robot.

Control unit The vital part of a robot that tells it what to do. This is normally a computer.

Data A collection of facts.

Data processing The conversion of information by a computer or similar device into more useful information.

Degrees of freedom The number of planes in which a robot arm, or parts of a robot arm, ca move. Most robots have six degrees of freedor Humans are rather better and have about thirt

Feedback The mechanism by which sequence of events influences action. A goc example of a device providing feedback is thermostat. Thermostats respond to heat and are often used to control the temperature central-heating systems automatically.

First-generation robot The most commc kind of robot in which a series of instructio controls the robot's arm. There is nothing fanc about these robots; they lack any "sense".

Flexible manufacturing systems A seri of robots and machine tools that turn out a lot different types of parts under computer contro

Hand The important part of a robot arm. N robot could work without one. Robot hanc include suction pads, claws and grippers.

Hydraulic robot A machine in which tl hand and arm of the robot are moved b pressurized liquid acting on mechanic linkages.

Knowledge base A feature of advanced form of computers which contains general data abou the world and feeds this to the rest of th computer.

Machine tool An industrial device that cut grinds, bores or otherwise fashions pieces metal, wood or plastic. The latest machine too are controlled by computers and are fed wit parts by robots.

Mechanization The first stage in the proce by which machines take over the work humans. This word first came into widesprea use with the "factory age" ushered in by th ndustrial Revolution.

Iemory The part of a computer used to hold r "store" programs and data.

'umerical control The process by which nachine tools or robots are controlled by binary igits.

'neumatic robot A robot which is driven by ne movement of compressed air.

'rogram Sequence of instructions that conrols what a computer does. All robots are ontrolled by computers.

Robot The modern definition is a mechanical rm controlled by a computer so it can copy the ctions of humans in a flexible, universal way. However, the term can loosely be taken to mean ny machine which copies human beings in ppearance or action. The word 'robot' comes rom the Czech word *robota*, which means servitude'.

Second-generation robot A robot with one r more sensors.

Sensor A device that feeds information from he surroundings to a person or device. Important human sensors are the eyes, ears, nose and nerve endings. Common types of sensors in obots are TV cameras and force- or pressure-ensors.

Space-robot A robot device specially built to unction outside the Earth's atmosphere.

Teaching by doing A method of instructing obots in which a human operator takes the obot through a particular action. The robot emembers the action for later use.

Telechiric arm Mechanical arm controlled continuously by a human operator. Not to be confused with robots, which are controlled by computers.

Terminal Unit connected to a computer to feed in, or gain access to, information. Terminals include keyboards and visual display units (screens).

Third-generation robot A robot with artificial intelligence.

Unimate A popular form of robot made by the American firm Unimation.

Unmanned factory An advanced form of factory. In reality a few people are required for jobs such as administration, maintenance, and controlling the computers. Unmanned plants will most probably become widespread in the 21st century.

Very large scale integration The process of packing many electronic components into a small space to make powerful computers.

Vision robot A robot with a television camera to enable it to "see" what it is doing.

Visual Display Unit (VDU) A cathode ray tube, as found in a TV set, for displaying the output of a computer.

Index

Note: Page numbers in *italics* refer to illustrations.

ACKNOWLEDGEMENTS

Front cover: bottom left, Clayton Bailey; bottom right, Cincinnati Milacron; contents page: Hitachi Ltd; 12: Cincinnati Milacron; 13 left: Trallfa Nils Underhang A/S; 13 right: British Robot Association; 14: UK Atomic Energy Authority; 16: British Leyland; 19: Devilbiss Co. Ltd; 20 bottom: Autoplace Inc./Visual Arts; 21 right: Hitachi Ltd; 22: UK Atomic Energy Authority; 24–25: NASA; 26: National Film Archive; 27 top left: BBC Copyright Photograph; 27 bottom left: Courtesy of Lucasfilm UK Ltd; 27 right: Clayton Bailey; 28; Dept of the United States Navy.

Picture research: Jackie Cookson

The publishers would like to thank the following for their help in the preparation of this book: British Robot Association; Taylor Hitec Ltd; University of Warwick Robot Laboratory.